Health 56

无需开采的金属
Metals without Mining

Gunter Pauli

冈特·鲍利 著

李欢欢 译

学林出版社
www.xuelinpress.com

丛书编委会

主　任：贾　峰

副主任：何家振　郑立明

委　员：牛玲娟　李原原　李曙东　吴建民　彭　勇
　　　　冯　缨　靳增江

丛书出版委员会

主　任：段学俭

副主任：匡志强　张　蓉

成　员：叶　刚　李晓梅　魏　来　徐雅清　田振军
　　　　蔡雩奇

特别感谢以下热心人士对译稿润色工作的支持：

姜竹青　韩　笑　杨　爽　周依奇　于　哲　阳平坚
李雪红　汪　楠　单　威　查振旺　李海红　姚爱静
朱　国　彭　江　于洪英　隋淑光　严　岷

目录

无需开采的金属	4
你知道吗？	22
想一想	26
自己动手！	27
学科知识	28
情感智慧	29
艺术	29
思维拓展	30
动手能力	30
故事灵感来自	31

Contents

Metals without Mining	4
Did you know?	22
Think about it	26
Do it yourself!	27
Academic Knowledge	28
Emotional Intelligence	29
The Arts	29
Systems: Making the Connections	30
Capacity to Implement	30
This fable is inspired by	31

细菌们聚在一起参加每周的聚会。和往常一样,他们嘲笑着人类。

"你们知道吗?他们用炸药来炸开地壳。"一个细菌说道。

The bacteria are gathering for their weekly get-together. As usual, they are poking fun at humans.

"Do you know they are using dynamite to break open the Earth's crust?" says one bacterium.

细菌们聚在一起参加每周的聚会

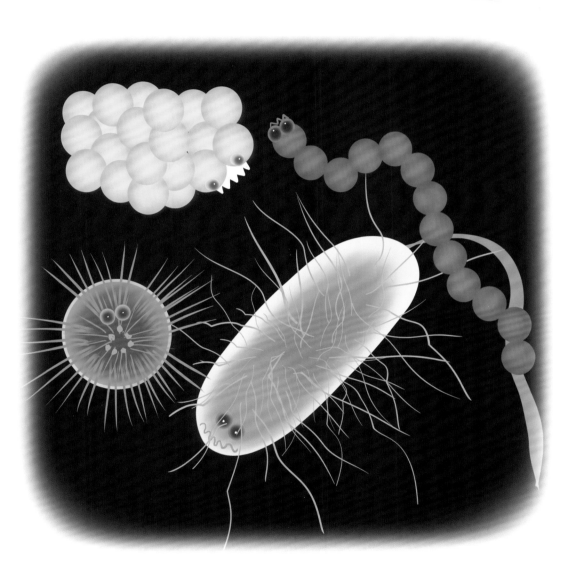

The bacteria are gathering for their weekly get-together

在地面上炸出大洞

Big holes in the ground

"他们有什么权利这么做?他们发现了火,就开始焚烧一切他们碰得到的东西,现在他们想毁灭一切吗?"

"有些人用炸药捕鱼,毁坏珊瑚,但大部分人用炸药在地面上炸出大洞。"

"What gives them the right to do that? They invented fire and started burning everything they could get their hands on, and now they want to blow everything up?"

"Some humans use dynamite to catch fish and destroy corals, but most use it to blow big holes in the ground."

"我想知道，"另一个细菌说道，"他们从来没和地衣——我们的专业矿工聊过吗？"

"想也别想！大多数人根本不知道地衣是什么！他们误以为它是植物，事实上，它是菌类和藻类的共生联合体，给双方都带来了巨大利益。"

"I wonder," says another bacterium. "Have they never talked to the lichens, our professional miners?"

"No ways! Most humans don't even know what lichen is! They mistake it for a plant, when really it's the friendship between mushroom and algae that brings great benefits to both."

和地衣聊过

Talked to the lichens

……最沉默的开采方式

...most silent type of mining

"人类如此无知,太丢人了!地衣能在岩石上挖出小沟,一天能挖几百米,然后把矿物质输送给植物。"

"他们从没认识到地衣只有两个细胞的厚度。这是最温和最沉默的开采方式。想想吧,如果用挖掘机而不是手术刀做胃部手术——光光这个想法就足以吓跑他们了!"

一个细菌咯咯地笑起来。

"These humans are so ignorant, it's embarrassing! Lichens can dig tiny tunnels in rocks, hundreds of metres a day, and deliver minerals to plants."

"They've never figured out that lichens are only two cells thick. It's the most benign and silent type of mining. Imagine proposing surgery of the stomach with an excavator instead of a scalpel – the idea alone will scare them off!" giggles a bacterium.

"听着，人类本来也可以成为温和的开采者。问题是他们根本不知道自己擅长什么。"

"你的意思是人类也有擅长的东西？别忘了，我们称他们为'无智慧的人'！"细菌们大笑起来。

"Look, humans could also be benign miners. The problem is they have never realised what they are good at."

"You mean humans are good at something?Remember, we call them the homo non sapiens!" The bacteria burst out laughing.

无智慧的人

Homo non sapiens

富含铁质的食物而不是垃圾食品

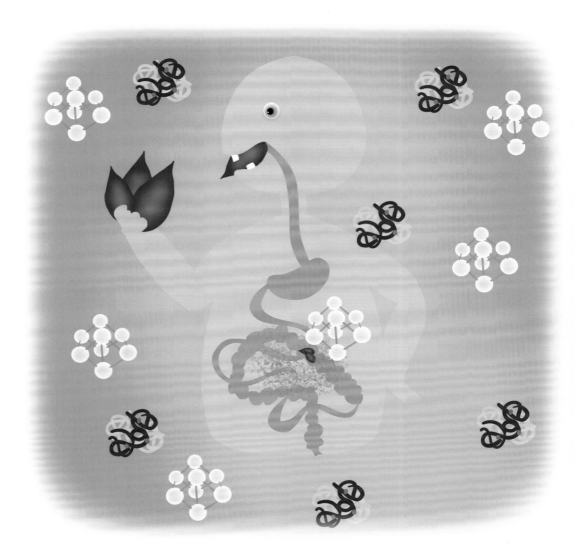

Iron-rich food instead of junk food

"听着,他们的祖先可没这么蠢,"一个年长的细菌说道。"他们吃富含铁质的食物,而不是现在的垃圾食物。他们的身体可以吸收少量铁元素,使其和氢、碳、氮和氧相结合,形成血红蛋白。"

"血红什么?哦,别讲了。这是晚会,不是化学课!"

"不,说真的。就像我们细菌可以从一堆垃圾中辨别和吸收铜、金、银或锌,人类可以利用铁元素做同样的事情。"

"Look, their ancestors weren't so dumb," says an older bacterium. "They ate iron-rich food instead of the junk food of today. And their bodies can take a minute amount of iron and mix it with hydrogen, carbon, nitrogen and oxygen to make haemoglobin."

"Haemo– what? Oh, stop it. This is an evening out, not a chemistry lesson!"

"No, really. Just as our bacteria friends can identify and absorb copper, gold, silver or zinc from a pile of dust, people can do the same with iron."

一些细菌对此提起了兴趣。"那为什么人类的血液需要铁元素呢?"

"嗯,铁元素帮助血液接收和输送更多的氧气,这样会使身体充满活力,也能让大脑更好地运转。"

Some of the bacteria are becoming interested. "So why do humans want iron in their blood?"

"Well, iron helps blood to pick up and transport more oxygen, which in turn energises the whole body, and makes the brain work better."

让大脑更好地运转

Makes the brain work better

如果他们真的如此聪明，干吗要……

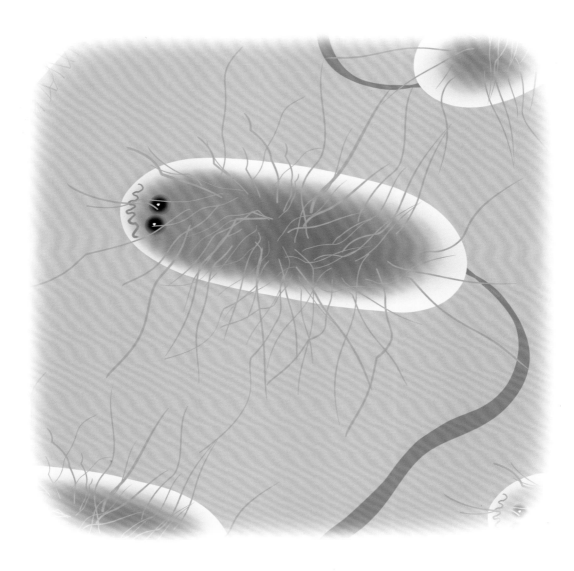

If humans are that intelligent, how come...

"对于人类来说是个不错的设计！"年长的细菌喃喃地说。

可是一个细菌急于发问。"如果他们真的如此聪明，干吗要用炸药把地球炸出巨大的坑？他们为什么要破坏海洋？又是谁提议把金属转化成气体？"

"Not a bad design for a human being!" mumbles the old one.

But one of the bacteria is eager to ask a question. "If they are that intelligent, how come they use dynamite to blow huge holes in the Earth? And why do they destroy the sea? And whoever proposed turning metals into gas?"

"我想知道，"又一个细菌开口道。"人类是愚蠢吗，或者他们只是无知？他们丢弃的电话和电脑里所含的金和其他金属，比他们能从地球开采的还多，他们没有意识到吗?"

"嗯，"另一个细菌若有所思地说，"他们确实行事鲁莽，就像闯进了瓷器店的牛。"

……这仅仅是开始！……

"I wonder," says another. "Are people stupid, or are they just ignorant? Don't they realise that they already have more gold and other metals in the phones and computers that they dump, than could ever be extracted from the Earth?"

"Hmm," another bacterium muses. "They certainly do behave like a bull in a china shop."

... AND IT HAS ONLY JUST BEGUN!...

……这仅仅是开始!……

…AND IT HAS ONLY JUST BEGUN!…

Did You Know? 你知道吗？

Dynamite is destructive as an explosive, but its main ingredient, nitroglycerine, is also used as a medicine to treat heart conditions.

甘油炸药是一种极具破坏性的炸药，不过甘油炸药的主要原料硝酸甘油也是一种治疗心脏病的药物。

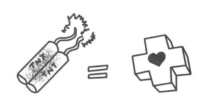

地衣由藻类和真菌共同组成：两者相互供给养料。

Lichens are a combination of algae and fungi: one makes food for the other.

Lichens inhibit the growth of ringworms and tuberculosis. Roccella lichen produces litmus, which is the standard method to measure pH.

地衣能抑制皮癣和结核病。从染料衣属地衣中可以提取石蕊，它是测量溶液酸碱度（pH 值）的标准方法。

Lichens are so strong, they can pulverise rock solely with the water pressure in their cells.

地衣非常强大，其细胞内的水压可以压碎岩石。

Iron, – a mineral which is abundant in meat, tofu and spinach – is a component of haemoglobin, the substance in red blood cells that carries oxygen through the body.

铁是肉类、豆腐和菠菜中富含的一种矿物质，是血红蛋白的组成成分。血红蛋白是红血球内的一种物质，负责将氧气输送到全身。

Chlorophyll is exactly the same molecule as haemoglobin, but with one atom difference: the iron is replaced by magnesium.

叶绿素和血红蛋白是几乎完全一样的分子，但有一个原子不同：镁代替了铁。

$\mathcal{B}$acteria and their synthetic counterparts can remove heavy metals from our bodies and can separate platinum group metals from ore.

细菌和细菌合成物能清除人体内的重金属，还能从矿石中分离出铂族金属。

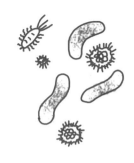

$\mathcal{J}$unk food has little nutritional value and offers high fat, sugar, salt and calories.

垃圾食品几乎没有营养价值，而且脂肪、糖、盐和热量含量都很高。

Think About It  想一想

Can you manufacture weapons while also honouring peace?

你能一边崇尚和平，一边制造武器吗？

有没有一种威力特别强大的武器以致于人们害怕使用它？

Could a weapon ever be so powerful that people would be afraid to use it?

How can a tiny alga supported by a small fungus be strong enough to crush a rock to pieces?

在微小的真菌帮助下，微小的藻类如何强到足以研碎岩石？

How is it possible that something bad can also be good?

有些坏事也可能又是好事，这怎么可能呢？

Do It Yourself !
自己动手！

Have a good look around your bedroom. How many products do you own that have been made using dynamite or other explosives? Make a list of everything you suspect is a result of the use of explosives. Present your list to your friends and family and, if you are ready, state that you do not want the face of the arth to be destroyed. Will anyone listen to you?

仔细查看你的卧室。你有多少件物品在生产过程中用了甘油炸药或其他炸药？将所有你怀疑使用过炸药的物品列成一个清单。让你的朋友和家人看看你的清单，如果你准备好了，请告诉他们你不希望地球表面遭到破坏。会有人听你的吗？

TEACHER AND PARENT GUIDE

学科知识
Academic Knowledge

生物学	地衣是混合型微生物，以藻类为基础，生存于真菌的藻丝中，这种共生关系出现于6亿年前，两种微生物都从中受益；组成地衣的藻类可以独立生长，或和真菌共同生长，但组成地衣的真菌只能和藻类共同存在；地衣覆盖了8%的地球表面，大部分分布于北极冻原；富含铁质的食物促进红血球生成。
化学	安全炸药的基础是硝化甘油；医学上，硝化甘油用于治疗心脏病和前列腺癌；石蕊是一种从地衣提取的染料混合物，渗入纸张制成试纸来指示pH值：酸性环境下，石蕊纸由蓝色转为红色，碱性环境下，红色石蕊纸又转为蓝色；如果缺铁，人体不能制造血红蛋白，因而患上贫血；螯合作用是细菌对特定的正离子有亲和力，借此有选择地离析金属物质。
物理	硝化甘油不耐冲击，因此需要土壤或木屑吸收；安全炸药捕鱼，即用炸药炸晕或炸死鱼群，以便捞捕，这样毁坏了环境，尤其是珊瑚礁；在潮湿的情况下，地衣能逐渐分解岩石。
工程学	在德国、意大利和瑞士，地衣用作空气污染的生物监测器；医学手术已经从能让手伸进人体做手术的切口，发展到显微镜和精密仪器的微创切口；通过高温蒸发，合金能重新利用。
经济学	本地天然的原料越来越少，取而代之的是来自世界各地的合成原料。
伦理学	你怎么能一边呼吁和平一边支持武器工业呢？在有其他方法可供选择时，采矿和矿石加工怎能继续依赖破坏性的做法呢？
历史	诺贝尔从克里米亚半岛油田赚到的钱比他的安全炸药专利要多。
地理	地衣到处都有，从雨林到沙漠，只有深海没有；在一些不适宜居住的环境下，地衣可能是唯一出现的植被。
数学	含有75%硝化甘油的0.19千克柱状安全炸药包大概含有1兆焦耳能量，相当于0.8千克TNT炸药的威力。
生活方式	石耳属地衣是美味佳肴，英语中叫岩石内脏，日本叫岩茸，韩国和中国叫石耳。
社会学	诺贝尔曾希望能生产一种极具毁灭性的材料，通过其威慑作用阻止战争发生。
心理学	幽默在批判中能否发挥作用？习语能更清楚巧妙地表述某事吗？
系统论	地衣和细菌等自然系统能开采岩石，而没有炸药毁坏性的副作用。

教师与家长指南

情感智慧
Emotional Intelligence

细菌

细菌热爱生活，乐于参加晚会，爱开玩笑。他们有批判性的眼光，展示了逻辑性来支持自己的观点。细菌们联系之前人类滥用火石、毁坏珊瑚的经历，没有给人类狡辩的权利。细菌们自娱自乐时，展现了学问、严谨和对技术细节的精通。他们有时间和位置感，利用对比来澄清争议。人类自称为智者，对此细菌们进行了批判，他们笑称人类为"无智慧的人"。当他们意识到人类的优点和弱点后，话题引向了解决方案，细菌们提出了一些深远的问题，但没有立刻得到回复。细菌们用一个习语进行总结，使得其语言听起来意思明确、说法聪明而且减轻了攻击性。

艺术
The Arts

怎么模仿爆炸声？让我们试着用嗓子。首先，自己想象一下爆炸听起来应该是怎么样的，然后大家一起喊出来。之后，每个人再次喊出自己认为的爆炸声。你注意到听到别人的声音后，自己的声音是怎么改变的吗？大家能发出一样的声音吗？当然，不可能每个人都发出一样的声音，尽管我们试着发出一样的声音，因为每个人的表达和艺术创造力不同。

TEACHER AND PARENT GUIDE

思维拓展
Systems: Making the Connections

　　采矿是一个具有破坏性的行业。几千年来实践证明，一旦矿产挖空了，矿区就成了鬼城，只有极少数例外。更糟的是，只重视利润的公司往往不负责任：他们可能破产倒闭，留下一片惨遭破坏的环境。对高产量的追求导致使用炸药。用于提取矿物和稀有金属的采矿技术极其依赖资源。不断地追求产量和经济规模，采矿过程用的人力越来越少，导致经济崩溃。环境恶化导致生态系统崩溃，环境污染影响着附近居民的健康和生活。如今，最大的金属来源是废弃的电子和医疗设备。但回收利用旧电子产品需要和采矿业相同的技术，同时极其依赖能源，却并没有创造更多的工作，因此不是非走那条路不可——还有其他可选的技术。工业过程自始至终都需要大量能源和原料，所以不会考虑小规模加工的商业模式。但如果我们能大大减少全球矿石和废弃物的运输，那就能证明其他技术也可以是有效的。无论如何，利用细菌对特定金属和矿物质的吸附力已成功开辟了一个市场。只有办公复印机大小的螯合机器每天能处理100到500部手机，以细微粉末形式回收金属，再循环利用。这为都市采矿开创了一条大道：有效利用资源，把危险废弃物转换成经济增长点。

动手能力
Capacity to Implement

　　你家附近有潜在的矿山吗？不要想着大矿山，想一想大量金属在哪里——也许就在你家里！金属无处不在：电线、冰箱电路板、电脑、打印机、电视、手机和玩具……到处都有！想象这是你的潜在矿山，开始策划一个当地的回收公司吧。想象有一台办公复印机大小的螯合机器，想一想你需要多少电子废弃物才能让公司运转起来。这不是写商业计划，而是设计一个商业模式。你会惊奇地发现原来身边有那么多金属，又有那么多金属闲置在家，而没能给你或社区带来收益。

教师与家长指南

故事灵感来自

亨利·科勒辛斯基
Henry Kolesinski

亨利·科勒辛斯基被朋友们称为汉克，他获得了位于美国马萨诸塞州波士顿的东北大学的化学专业学位。汉克有多年合成化学方面的背景。在拍立得公司，他开发了大量用于摄影行业的特殊化学品，并获得专利。担任密理博沃特公司的咨询师期间，他获得了生命科学的经历。他创办了团结科技公司，提供色谱仪，该企业后被热电公司并购。

他有多项获批的专利和尚待批准的专利，发表了多篇技术论文。汉克和罗伯特·科里一起创立了 Prime Separations 公司，生产分子分离和分子净化机器，这些分子既可以是微小的有机化合物、稀有金属，也可以是蛋白质和污染物质。

更多资讯

http://primeseparations.com

图书在版编目（CIP）数据

无需开采的金属：汉英对照 /（比）鲍利著；李欢欢译. —— 上海：学林出版社，2015.6
（冈特生态童书. 第 2 辑）
ISBN 978-7-5486-0850-9

Ⅰ. ①无… Ⅱ. ①鲍… ②李… Ⅲ. ①生态环境－环境保护－儿童读物－汉、英 Ⅳ. ① X171.1-49

中国版本图书馆 CIP 数据核字（2015）第 086065 号

————————————————————————

© 2015 Gunter Pauli
著作权合同登记号 图字 09-2015-446 号

冈特生态童书
无需开采的金属

作　　者——	冈特·鲍利
译　　者——	李欢欢
策　　划——	匡志强
责任编辑——	李晓梅
装帧设计——	魏　来
出　　版——	上海世纪出版股份有限公司 学林出版社
	地　址：上海钦州南路 81 号　　电话/传真：021-64515005
	网址：www.xuelinpress.com
发　　行——	上海世纪出版股份有限公司发行中心
	（上海福建中路 193 号 网址：www.ewen.co）
印　　刷——	上海图宇印刷有限公司
开　　本——	710×1020　1/16
印　　张——	2
字　　数——	5 万
版　　次——	2015 年 6 月第 1 版
	2015 年 6 月第 1 次印刷
书　　号——	ISBN 978-7-5486-0850-9/G·299
定　　价——	10.00 元

（如发生印刷、装订质量问题，读者可向工厂调换）